YOUR KNOWLEDGE HAS VALUE

- We will publish your bachelor's and master's thesis, essays and papers

- Your own eBook and book - sold worldwide in all relevant shops

- Earn money with each sale

Upload your text at www.GRIN.com and publish for free

Iduabo John Afa

A Study on Liquid Crystal Display (LCD) in Optoelectronics

GRIN Verlag

Bibliografische Information der Deutschen Nationalbibliothek:

Die Deutsche Bibliothek verzeichnet diese Publikation in der Deutschen National-
bibliografie; detaillierte bibliografische Daten sind im Internet über http://dnb.d-
nb.de/ abrufbar.

Dieses Werk sowie alle darin enthaltenen einzelnen Beiträge und Abbildungen
sind urheberrechtlich geschützt. Jede Verwertung, die nicht ausdrücklich vom
Urheberrechtsschutz zugelassen ist, bedarf der vorherigen Zustimmung des Verla-
ges. Das gilt insbesondere für Vervielfältigungen, Bearbeitungen, Übersetzungen,
Mikroverfilmungen, Auswertungen durch Datenbanken und für die Einspeicherung
und Verarbeitung in elektronische Systeme. Alle Rechte, auch die des auszugsweisen
Nachdrucks, der fotomechanischen Wiedergabe (einschließlich Mikrokopie) sowie
der Auswertung durch Datenbanken oder ähnliche Einrichtungen, vorbehalten.

Imprint:

Copyright © 2011 GRIN Verlag GmbH
Druck und Bindung: Books on Demand GmbH, Norderstedt Germany
ISBN: 978-3-656-41917-4

This book at GRIN:

http://www.grin.com/en/e-book/213415/a-study-on-liquid-crystal-display-lcd-in-
optoelectronics

GRIN - Your knowledge has value

Der GRIN Verlag publiziert seit 1998 wissenschaftliche Arbeiten von Studenten, Hochschullehrern und anderen Akademikern als eBook und gedrucktes Buch. Die Verlagswebsite www.grin.com ist die ideale Plattform zur Veröffentlichung von Hausarbeiten, Abschlussarbeiten, wissenschaftlichen Aufsätzen, Dissertationen und Fachbüchern.

Visit us on the internet:

http://www.grin.com/

http://www.facebook.com/grincom

http://www.twitter.com/grin_com

A STUDY ON LIQUID CRYSTAL DISPLAY (LCD) IN OPTO-ELECTRONICS

UG/07/1657
Afa Iduabo John
Department of Physics
Niger Delta University
PMB 71, Wilberforce Island, Bayelsa State

Abstract:

Liquid crystals are understood not to emit light directly. The idea of liquid crystal display (LCD) is that they use the light modulating properties of liquid crystals. These LCDs are used in a wide range of applications including computer monitors, gaming devices, video players, watches, clock display, calculators and many more.
The aim of our study is to show how with the use of the concept of LCs, LCDs have replaced older display methods such as Cathode Ray tube displays in display in modern devices like computer monitor display. We talk about the history, quality control, classifications and uses of LCDs. The advantages of these LCDs have proved to be far more over the CRTs. It is concluded that LCDs are currently the best for monitor and screen applications.

Keywords: liquid crystals, electronic visual display, polarizing filter film (Polarization filter), pixels, bistable LCD.

Literature Review

Liquid crystals (LCs) are a state of matter that have properties between those of a conventional liquid and those of a solid crystal (Chandrasekhar, 1994) For instance, an LC may flow like a liquid, but its molecules may be oriented in a crystal-like way. There are many different types of LC phases, which can be distinguished by their different optical properties. When viewed under a microscope using a polarized light source, different liquid crystal phases will appear to have distinct textures. The contrasting areas in the textures correspond to domains where the LC molecules are oriented in different directions. Liquid crystals can be divided into thermotropic, lyotropic and metallotropic phases.

Liquid crystals find wide use in liquid crystal displays, which rely on the optical properties of certain liquid crystalline substances in the presence or absence of an electric field. In a typical device, a liquid crystal layer (typically 10 μm thick) sits between two polarizers that are crossed (oriented at 90° to one another). The liquid crystal alignment is chosen so that its relaxed phase is a twisted one (Castellano, 2005). This twisted phase reorients light that has passed through the first polarizer, allowing its transmission through the second polarizer (and reflected back to the observer if a reflector is provided). The device thus appears transparent. When an electric field is applied to the LC layer, the long molecular axes tend to align parallel to the electric field thus gradually untwisting in the center of the liquid crystal layer. In this state, the LC molecules do not reorient light, so the light polarized at the first polarizer is absorbed at the second polarizer, and the device loses transparency with increasing voltage. In this way, the electric field can be

used to make a pixel switch between transparent or opaque on command. Color LCD systems use the same technique, with color filters used to generate red, green, and blue pixels (Castellano, 2005). Similar principles can be used to make other liquid crystal based optical devices (Alkeskjold *et al.*, 2007).

1.0 Introduction

The idea of application of liquid crystals brings us to our study of liquid crystal display (LCD). Each pixel of an LCD typically consists of a layer of molecules aligned between two transparent electrodes, and two polarizing filters, the axes of transmission of which are (in most of the cases) perpendicular to each other. With no actual liquid crystal between the polarizing filters, light passing through the first filter would be blocked by the second (crossed) polarizer. In most of the cases the liquid crystal has double refraction (Collings and Hird, 1997).

The surfaces of the electrodes that are in contact with the liquid crystal material are treated so as to align the liquid crystal molecules in a particular direction. This treatment typically consists of a thin polymer layer that is unidirectionally rubbed using, for example, a cloth. The direction of the liquid crystal alignment is then defined by the direction of rubbing. Electrodes are made of a transparent conductor called Indium Tin Oxide (ITO). Fig 1 shows the structure of a Liquid Crystal Display (retrieved from Wikipedia.org)

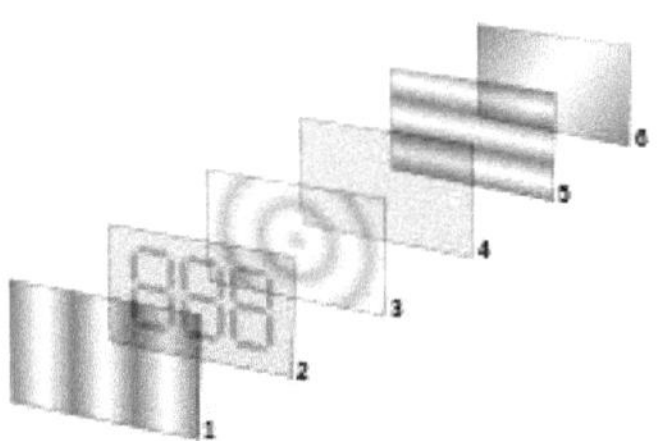

Layers: 1 – vertical polarization filter, 2,4 – glass with electrodes (ITO), 3 – liquid crystals, 5 – horizontal polarization filter, 6 – reflector

Fig 1: Structure of liquid crystal display

Before applying an electric field, the orientation of the liquid crystal molecules is determined by the alignment at the surfaces of electrodes. In a twisted nematic device (still the most common liquid crystal device), the surface alignment directions at the two electrodes are perpendicular to each other, and so the molecules arrange themselves in a helical structure, or twist. This reduces the rotation of the polarization of the incident light, and the device appears grey. If the applied voltage is large enough, the liquid crystal molecules in the center of the layer are almost completely untwisted and the polarization of the incident light is not rotated as it passes through the liquid crystal layer. This light will then be mainly polarized perpendicular to the second filter, and thus be blocked and the pixel will appear black. By controlling the voltage applied across the liquid crystal layer in each pixel, light can be allowed to pass through in varying

amounts thus constituting different levels of gray. This electric field also controls (reduces) the double refraction properties of the liquid crystal (Collings and Hird, 1997)

The optical effect of a twisted nematic device in the voltage-on state is far less dependent on variations in the device thickness than that in the voltage-off state. Because of this, these devices are usually operated between crossed polarizers such that they appear bright with no voltage (the eye is much more sensitive to variations in the dark state than the bright state). These devices can also be operated between parallel polarizers, in which case the bright and dark states are reversed. The voltage-off dark state in this configuration appears blotchy, however, because of small variations of thickness across the device.
Both the liquid crystal material and the alignment layer material contain ionic compounds. If an electric field of one particular polarity is applied for a long period of time, this ionic material is attracted to the surfaces and degrades the device performance. This is avoided either by applying an alternating current or by reversing the polarity of the electric field as the device is addressed (the response of the liquid crystal layer is identical, regardless of the polarity of the applied field).

1.1 Construction

As shown in fig 2 (a), a liquid crystal "cell" consists of a thin layer (about 10μm) of a liquid crystal sandwiched between two glass sheets with transparent electrodes deposited on their inside faces. With both glass sheets transparent, the cell is known as *transmittive type* cell. When one glass is transparent and the other has a reflective coating, the cell is called *reflective coating*, the cell is called *reflective type*. The LCD does not produce any illumination of its own. It, in fact, depends entirely on illumination falling on it from an external source for its visual effect.

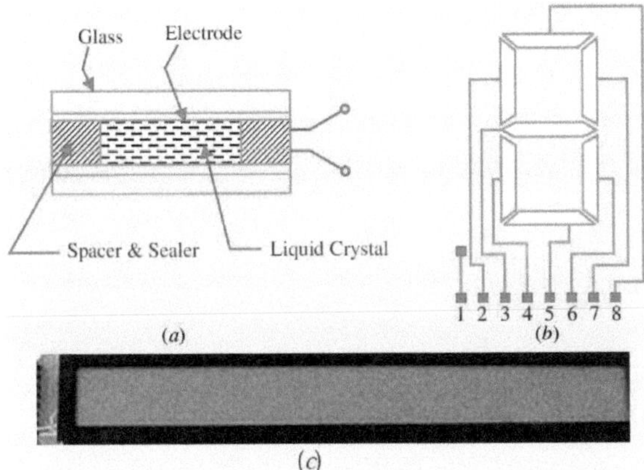

Fig. 2(a-c): Picture of Liquid crystal cell, a digit on an LCD, and an LCD used in portable instrument respectively.

1.2 Working

The two types of display available are known as (i) field-effect display and (ii) dynamic scattering display. When field-effect display is energized, the energized areas of the LCD absorb the incident light and, hence give localized black display. When dynamic scattering display is energized, the molecules of energized area of display become turbulent and scatter light in all directions. Consequently, the activated areas take on a frosted glass appearance resulting in a silver display. Of course, the unenergized areas remain translucent.
As shown in fig 2 (b), a digit on an LCD has a segment appearance. For example, if number 5 is required, the terminals 8, 2, 3, 6 and 5 would be energized so that only these regions would be activated while the other areas would remain clear. Fig 2 (c) shows the picture of an LCD used in portable instrument.

2.0 Brief History:

- 1888: Friedrich Reinitzer (1858–1927) discovers the liquid crystalline nature of cholesterol extracted from carrots (that is, two melting points and generation of colours) and published his findings at a meeting of the Vienna Chemical Society on May 3, 1888 (F. Reinitzer: *Beiträge zur Kenntniss des Cholesterins, Monatshefte für Chemie (Wien) 9, 421-441 (1888)*).

- 1904: Otto Lehmann publishes his work *"Flüssige Kristalle"* (Liquid Crystals).

- 1911: Charles Mauguin first experiments of liquids crystals confined between plates in thin layers.

- 1922: Georges Friedel describes the structure and properties of liquid crystals and classified them in 3 types (nematics, smectics and cholesterics).

- 1927: Vsevolod Frederiks devises the electrically switched light valve, called the Fréedericksz transition, the essential effect of all LCD technology.

- 1936: The Marconi Wireless Telegraph company patents the first practical application of the technology, *"The Liquid Crystal Light Valve"*.

- 1962: The first major English language publication on the subject *"Molecular Structure and Properties of Liquid Crystals"*, by Dr. George W. Gray.

- 1962: Richard Williams of RCA found that liquid crystals had some interesting electro-optic characteristics and he realized an electro-optical effect by generating stripe-patterns in a thin layer of liquid crystal material by the application of a voltage. This effect is based on an electro-hydrodynamic instability forming what is now called "Williams domains" inside the liquid crystal.

- 1964: George H. Heilmeier, then working in the RCA laboratories on the effect discovered by Williams achieved the switching of colours by field-induced realignment of dichroic dyes in a homeotropically oriented liquid crystal. Practical problems with this new electro-optical effect made Heilmeier continue to work on scattering effects in liquid crystals and finally the achievement of the first operational liquid crystal display based on what he called the *dynamic scattering mode* (DSM). Application of a voltage to a DSM display switches the initially clear transparent liquid crystal layer into a milky turbid state. DSM displays could be operated in transmissive and in reflective mode but they required a considerable current to flow for their operation.(Castellano, 2006) George H. Heilmeier was inducted in the National Inventors Hall of Fame and credited with the invention of LCD.

- 1960s: Pioneering work on liquid crystals was undertaken in the late 1960s by the UK's Royal Radar Establishment at Malvern, England. The team at RRE supported ongoing work by George Gray and his team at the University of Hull who ultimately discovered the cyanobiphenyl liquid crystals (which had correct stability and temperature properties for application in LCDs).

- 1970: On December 4, 1970, the twisted nematic field effect in liquid crystals was filed for patent by Hoffmann-LaRoche in Switzerland, (Swiss patent No. 532 261) with Wolfgang Helfrich and Martin Schadt (then working for the Central Research Laboratories) listed as inventors. Hoffmann-La Roche then licensed the invention to the Swiss manufacturer Brown, Boveri & Cie who produced displays for wrist watches during the 1970s and also to Japanese electronics industry which soon produced the first digital quartz wrist watches with TN-LCDs and numerous other products. James Fergason while working with Sardari Arora and Alfred Saupe at Kent State University Liquid Crystal Institute filed an identical patent in the USA on April 22, 1971. In 1971 the company of Fergason ILIXCO (now LXD Incorporated) produced the first LCDs based on the TN-effect, which soon superseded the poor-quality DSM types due to improvements of lower operating voltages and lower power consumption.

- 1972: The first active-matrix liquid crystal display panel was produced in the United States by Westinghouse, in Pittsburgh, PA.

- 1996 Samsung develops the optical patterning technique that enables multi-domain LCD. Multi-domain and In Plane Switching subsequently remain the dominant LCD designs through 2010.

- 1997 Hitachi resurrects the In Plane Switching (IPS) technology producing the first LCD to have the visual quality acceptable for TV application.

- 2001 Jean Paul Gaultier uses LCD technology at his 2001 fall collection fashionshow which brings LCD to mainstream.

- 2007: In the 4Q of 2007 for the first time LCD televisions surpassed CRT units in worldwide sales.

- 2008: LCD TVs become the majority with a 50% market share of the 200 million TVs forecast to ship globally in 2008 according to Display Bank

2.1 Illumination

As LCD panels produce no light of their own, they require an external lighting mechanism to be easily visible. On most displays, this consists of a cold cathode fluorescent lamp that is situated behind the LCD panel. Passive-matrix displays are usually not backlit, but active-matrix displays almost always are, with a few exceptions such as the display in the original Gameboy Advance. Recently, two types of LED backlit LCD displays have appeared in some televisions as an alternative to conventional backlit LCDs. In one scheme, the LEDs are used to backlight the entire LCD panel. In another scheme, a set of red, green and blue LEDs is used to illuminate a small cluster of pixels, which can improve contrast and black level in some situations. For example, the LEDs in one section of the screen can be dimmed to produce a dark section of the image while the LEDs in another section are kept bright. Both schemes also allows for a slimmer panel than on conventional displays.

3.0 Passive and Active Matrix Addressed LCDs

Active-matrix addressed displays look "brighter" and "sharper" than passive-matrix addressed displays of the same size, and generally have quicker response times, producing much better images.

3.1 Passive Matrix addressed displays

LCDs with a small number of segments, such as those used in digital watches and pocket calculators, have individual electrical contacts for each segment. An external dedicated circuit supplies an electric charge to control each segment. This display structure is unwieldy for more than a few display elements.

Small monochrome displays such as those found in personal organizers, electronic weighing scales, older laptop screens, and the original Nintendo Game Boy have a passive-matrix structure employing super-twisted nematic (STN) (Gameboy user manual, 2011) or double-layer STN (DSTN) technology (the latter of which addresses a colour-shifting problem with the former), and colour-STN (CSTN) in which colour is added by using an internal filter. Each row or column of the display has a single electrical circuit. The pixels are addressed one at a time by row and column addresses. This type of display is called *passive-matrix addressed* because the pixel must retain its state between refreshes without the benefit of a steady electrical charge. As the number of pixels (and, correspondingly, columns and rows) increases, this type of display becomes less feasible. Very slow response times and poor contrast are typical of passive-matrix addressed LCDs.

Monochrome passive-matrix LCDs were standard in most early laptops (although a few used plasma displays).

3.2 Active Matrix addressed displays

The commercially unsuccessful Macintosh Portable (released in 1989) was one of the first to use an active-matrix display (though still monochrome), but passive-matrix was the norm until the

mid-1990s, when colour active-matrix became standard on all laptops. High-resolution colour displays such as modern LCD computer monitors and televisions use an active matrix structure. A matrix of thin-film transistors (TFTs) is added to the polarizing and colour filters. Each pixel has its own dedicated transistor, allowing each column line to access one pixel. When a row line is activated, all of the column lines are connected to a row of pixels and the correct voltage is driven onto all of the column lines. The row line is then deactivated and the next row line is activated. All of the row lines are activated in sequence during a refresh operation. Active-matrix addressed displays look "brighter" and "sharper" than passive-matrix addressed displays of the same size, and generally have quicker response times, producing much better images.

Types of Active Matrix Technology

3.2.1 Twisted nematic (TN)
Twisted nematic displays contain liquid crystal elements which twist and untwist at varying degrees to allow light to pass through. When no voltage is applied to a TN liquid crystal cell, the light is polarized to pass through the cell. In proportion to the voltage applied, the LC cells twist up to 90 degrees changing the polarization and blocking the light's path. By properly adjusting the level of the voltage almost any grey level or transmission can be achieved.

3.2.2 In-plane switching (IPS)
In-plane switching is an LCD technology which aligns the liquid crystal cells in a horizontal direction. In this method, the electrical field is applied through each end of the crystal, but this requires two transistors for each pixel instead of the single transistor needed for a standard thin-film transistor (TFT) display.

3.2.3 Advanced fringe field switching (AFFS)
Known as fringe field switching (FFS) until 2003 (Lee, *et al*, 2006), advanced fringe field switching is a technology similar to IPS or S-IPS offering superior performance and colour gamut with high luminosity. AFFS is developed by Hydis Technologies Co.,Ltd, Korea (Lee, *et al*, 2006).
AFFS-applied notebook applications minimize colour distortion while maintaining its superior wide viewing angle for a professional display. Colour shift and deviation caused by light leakage is corrected by optimizing the white gamut which also enhances white/grey reproduction.

3.2.4 Vertical alignment (VA)
Vertical alignment displays are a form of LCDs in which the liquid crystal material naturally exists in a vertical state removing the need for extra transistors (as in IPS). When no voltage is applied, the liquid crystal cell remains perpendicular to the substrate creating a black display. When voltage is applied, the liquid crystal cells shift to a horizontal position, parallel to the substrate, allowing light to pass through and create a white display. VA liquid crystal displays provide some of the same advantages as IPS panels.

3.2.5 Blue Phase mode
Blue phase mode LCDs have been shown as engineering samples early in 2008, they are not in mass-production yet. The physics of blue phase mode LCDs suggest that very short switching

times (~1 ms) can be achieved, so time sequential color control can possibly be realized and expensive color filters would be obsolete.

4.0 Quality control

Some LCD panels have defective transistors, causing permanently lit or unlit pixels which are commonly referred to as stuck pixels or dead pixels respectively. Unlike integrated circuits (ICs), LCD panels with a few defective transistors are usually still usable. It is claimed that it is economically prohibitive to discard a panel with just a few defective pixels because LCD panels are much larger than ICs, but this has never been proven. Manufacturers' policies for the acceptable number of defective pixels vary greatly. At one point, Samsung held a zero-tolerance policy for LCD monitors sold in Korea (Forbes.com, 2004). As of 2005, though, Samsung adheres to the less restrictive ISO 13406-2 standard (Samsung, 2005). Other companies have been known to tolerate as many as 11 dead pixels in their policies (Lenovo, 2007). Dead pixel policies are often hotly debated between manufacturers and customers. To regulate the acceptability of defects and to protect the end user, ISO released the ISO 13406-2 standard (Anders Jacobsen's blog, 2006). However, not every LCD manufacturer conforms to the ISO standard and the ISO standard is quite often interpreted in different ways.

LCD panels are more likely to have defects than most ICs due to their larger size. For example, a 300 mm SVGA LCD has 8 defects and a 150 mm wafer has only 3 defects. However, 134 of the 137 dies on the wafer will be acceptable, whereas rejection of the LCD panel would be a 0% yield. Due to competition between manufacturers quality control has been improved. An SVGA LCD panel with 4 defective pixels is usually considered defective and customers can request an exchange for a new one. Some manufacturers, notably in South Korea where some of the largest LCD panel manufacturers, such as LG, are located, now have "zero defective pixel guarantee", which is an extra screening process which can then determine "A" and "B" grade panels. Many manufacturers would replace a product even with one defective pixel. Even where such guarantees do not exist, the location of defective pixels is important. A display with only a few defective pixels may be unacceptable if the defective pixels are near each other. Manufacturers may also relax their replacement criteria when defective pixels are in the center of the viewing area. LCD panels also have defects known as *clouding* (or less commonly *mura*), which describes the uneven patches of changes in luminance. It is most visible in dark or black areas of displayed scenes.

5.0 LCD Classifications

There are various **classifications** of the electro-optical modes of **liquid crystal displays** (LCDs). We have; Polymer stabilized LCDs, Bistable LCDs (Zenithal and Azimuthal bistabilities), TN, VA and IPS-LCDs.

The operation of TN, VA and IPS-LCDs can be summarized as follows:

- a well aligned LC configuration is deformed by an applied electric field,
- this deformation changes the orientation of the local LC optical axis with respect to the direction of light propagation through the LC layer,
- this change of orientation changes the polarization state of the light propagating through the LC layer,

- this change of the polarization state is converted into a change of intensity by dichroic absorption, usually by external dichroic polarizers.

5.1 Zero-power (bistable) displays

The zenithal bistable device (ZBD), developed by QinetiQ (formerly DERA), can retain an image without power. The crystals may exist in one of two stable orientations ("Black" and "White") and power is only required to change the image. ZBD Displays is a spin-off company from QinetiQ who manufacture both grayscale and colour ZBD devices.
A French company, Nemoptic, has developed the BiNem zero-power, paper-like LCD technology which has been mass-produced in partnership with Seiko since 2007 (Nemoptic.com). This technology is intended for use in applications such as Electronic Shelf Labels, E-books, E-documents, E-newspapers, E-dictionaries, Industrial sensors, Ultra-Mobile PCs, etc.
In 2004 researchers at the University of Oxford demonstrated two new types of zero-power bistable LCDs based on Zenithal bistable techniques (Uche C., 2007).
Several bistable technologies, like the 360° BTN and the bistable cholesteric, depend mainly on the bulk properties of the liquid crystal (LC) and use standard strong anchoring, with alignment films and LC mixtures similar to the traditional monostable materials.

5.2 Specifications

Important factors to consider when evaluating a Liquid Crystal Display (LCD):

- **Resolution versus Range:** Fundamentally resolution is the granularity (or number of levels) with which a performance feature of the display is divided. Resolution is often confused with range or the total end-to-end output of the display. Each of the major features of a display has both a resolution and a range that are tied to each other but very different. Frequently the range is an inherent limitation of the display while the resolution is a function of the electronics that make the display work.

- **Spatial Performance:** LCDs come in only one size for a variety of applications and a variety of resolutions within each of those applications. LCD spatial performance is also sometimes described in terms of a "dot pitch". The size (or spatial range) of an LCD is always described in terms of the diagonal distance from one corner to its opposite. This is a historical aspect from the early days of CRT TV when CRT screens were manufactured on the bottoms of a glass bottle. The diameter of the bottle determined the size of the screen. Later, when TVs went to a more square format, the square screens were measured diagonally to compare with the older round screens.

The spatial resolution of an LCD is expressed in terms of the number of columns and rows of pixels (e.g., 1024×768). This had been one of the few features of LCD performance that was easily understood and not subject to interpretation. Each pixel is usually composed of a red, green, and blue sub pixel. However there are newer schemes to share sub-pixels among pixels and to add additional colours of sub-pixels. So going forward, spatial resolution may be more subject to interpretation.

One external factor to consider in evaluating display resolution is the resolution of your own eyes. For a normal person with 20/20 vision, the resolution of your eyes is about one minute of arc. In practical terms that means for an older standard definition TV set the ideal viewing distance was about 8 times the height (not diagonal) of the screen away. At that distance the individual rows of pixels merge into a solid. If you were closer to the screen than that, you would be able to see the individual rows of pixels. If you are further away, the image of the rows of pixels still merge, but the total image becomes smaller as you get further away. For an HDTV set with slightly more than twice the number of rows of pixels, the ideal viewing distance is about half what it is for a standard definition set. The higher the resolution, the closer you can sit to the set or the larger the set can usefully be sitting at the same distance as an older standard definition display.

For a computer monitor or some other LCD that is being viewed from a very close distance, resolution is often expressed in terms of dot pitch or pixels per inch. This is consistent with the printing industry (another form of a display). Magazines, and other premium printed media are often at 300 dots per inch. As with the distance discussion above, this provides a very solid looking and detailed image. LCDs, particularly on mobile devices, are frequently much less than this as the higher the dot pitch, the more optically inefficient the display and the more power it burns. Running the LCD is frequently half, or more, of the power consumed by a mobile device.

An additional consideration in spatial performance is viewing cone and aspect ratio. The **Aspect ratio** is the ratio of the width to the height (for example, 4:3, 5:4, and 16:9 or 16:10). Older, standard definition TVs were 4:3. Newer, HDTV's are 16:9 as are most new notebook computers. Movies are often filmed in much different (wider) aspect ratios which is why there will frequently still be black bars at the top and bottom of a HDTV screen.

The Viewing Angle of an LCD may be important depending on its use or location. The viewing angle is usually measured as the angle where the contrast of the LCD falls below 10:1. At this point, the colours usually start to change and can even invert, red becoming green and so forth. Viewing angles for LCDs used to be very restrictive however, improved optical films have been developed that give almost 180 degree viewing angles from left to right. Top to bottom viewing angles may still be restrictive, by design, as looking at an LCD from an extreme up or down angle is not a common usage model and these photons are wasted. Manufacturers commonly focus the light in a left to right plane to obtain a brighter image here.

- **Temporal/Timing Performance:** Contrary to spatial performance, temporal performance is a feature where smaller is better. Specifically, the range is the pixel response time of an LCD, or how quickly you can change a sub-pixel's brightness from one level to another. For LCD monitors, this is measured in btb (black to black) or gtg (gray to gray). These different types of measurements make comparison difficult. Further, this number is almost never published in sales advertising.

Refresh rate or the temporal resolution of an LCD is the number of times per second in which the display draws the data it is being given. Since activated LCD pixels do not flash on/off between frames, LCD monitors exhibit no refresh-induced flicker, no matter how low the refresh. High-end LCD televisions now feature up to 240 Hz refresh rate, which requires advanced digital processing to insert additional interpolated frames between the real images to smooth the image motion. However, such high refresh rates may not be actually supported by pixel response times and the result can be visual artifacts that distort the image in unpleasant ways. Temporal

performance can be further taxed if it is a 3D display. 3D displays work by showing a different series of images to each eye, alternating from eye to eye. For a 3D display it must display twice as many images in the same period of time as a conventional display and consequently the response time of the LCD becomes more important. 3D LCDs with marginal response times, will exhibit image smearing.

The temporal resolution of human perception is about 1/100th of a second. It is actually greater in your black and white vision (rod cells) than in colour vision (cone cells). You are more able to see flicker or any sort of temporal distortion in a display image by not looking directly at it as your rods are mostly grouped at the periphery of your vision.

- **Colour Performance:** There are many terms to describe colour performance of an LCD. They include colour gamut which is the range of colours that can be displayed and colour depth which is the colour resolution or the resolution or fineness with which the colour range is divided. Although colour gamut can be expressed as three pairs of numbers, the XY coordinates within colour space of the reddest red, greenest green, and bluest blue, it is usually expressed as a ratio of the total area within colour space that a display can show relative to some standard such as saying that a display was "120% of NTSC". NTSC is the National Television Standards Committee, the old standard definition TV specification. Colour gamut is a relatively straight forward feature. However with clever optical techniques that are based on the way humans see colour, termed colour stretch, colours can be shown that are outside of the nominal range of the display. In any case, colour range is rarely discussed as a feature of the display as LCDs are designed to match the colour ranges of the content that they are intended to show. Having a colour range that exceeds the content is a useless feature.

Colour Depth or colour support is sometimes expressed in bits, either as the number of bits per sub-pixel or the number of bits per pixel. This can be ambiguous as an 8-bit colour LCD can be 8 total bits spread between red, green, and blue or 8 bits each for each colour in a different display. Further, LCDs sometimes use a technique called dithering which is time averaging colours to get intermediate colours such as alternating between two different colours to get a colour in between. This doubles the number of colours that can be displayed; however this is done at the expense of the temporal performance of the display. Dithering is commonly used on computer displays where the images are mostly static and the temporal performance is unimportant.

When colour depth is reported as colour support, it is usually stated in terms of number of colours the LCD can show. The number of colours is the translation from the base 2-bit numbers into common base-10. For example, s 8-bit, in common terms means 2 to the 8th power or 256 colours. 8-bits per colour or 24-bits would be 256 x 256 x 256 or over 16 Million colours. The colour resolution of the human eye depends on both the range of colours being sliced and the number of slices; but for most common displays the limit is about 28-bit colour. LCD TVs commonly display more than that as the digital processing can introduce colour distortions and the additional levels of colour are needed to ensure true colours.

There are additional aspects to LCD colour and colour management such as white point and gamma correction which basically describe what colour white is and how the other colours are displayed relative to white. LCD televisions also frequently have facial recognition software which recognizes that an image on the screen is a face and both adjust the colour and the focus differently from the rest of the image. These adjustments can have important impact to the

consumer but are not easily quantifiable; people like what they like and everyone does not like the same thing. There is no substitute for looking at the LCD you are going to buy before buying it. Portrait film, another form of display, has similar adjustments built in to it. Many years ago, Kodak had to overcome initial rejection of its portrait film in Japan because of these adjustments. In the US, people generally prefer a more colour facial image than is reality (higher colour saturation). In Japan, consumers generally prefer a less saturated image. The film that Kodak initially sent to Japan was biased in exactly the wrong direction for Japanese consumers. TV sets have their built in biases as well.

- **Brightness and Contrast ratio:** Contrast ratio is the ratio of the brightness of a full-on pixel to a full-off pixel and, as such, would be directly tied to brightness if not for the invention of the blinking backlight (or burst dimming). The LCD itself is only a light valve, it does not generate light; the light comes from a backlight that is either a florescent tube or a set of LEDs. The blinking backlight was developed to improve the motion performance of LCDs by turning the backlight off while the liquid crystals were in transition from one image to another. However, a side benefit of the blinking backlight was infinite contrast. The contrast reported on most LCDs is what the LCD is qualified at, not its actual performance. In any case, there are two large caveats to contrast ratio as a measure of LCD performance.

The first caveat is that contrast ratios are measured in a completely dark room. In actual use, the room is never completely dark as you will always have the light from the LCD itself. Beyond that, there may be sunlight coming in through a window or other room lights that reflect off of the surface of the LCD and degrade the contrast. As a practical matter, the contrast of an LCD is governed by the amount of surface reflections not by the performance of the display.

The second caveat is that the human eye can only image a contrast ratio of a maximum of about 200:1. Black print on a white paper is about 15-20:1. That is why viewing angles are specified to the point where the fall below 10:1. A 10:1 image is not great, but is discernable.

Brightness is usually stated as the maximum output of the LCD. In the CRT era, Trinitron CRTs had a brightness advantage over the competition so brightness was commonly discussed in TV advertising. With current LCD technology, brightness, though important, is usually the same from maker to maker and is consequently not discussed much except for notebook LCDs and other displays that will be viewed in bright sunlight. In general, brighter is better but there is always a trade-off between brightness and battery life in a mobile device.

5.3 General Advantages and Uses of LCDs

An LCD has the distinct advantage of extremely low power requirement (about 10-15μW per 7-segment display as compared to a few mW for a LED). It is due to the fact that it does not itself generate any illumination but depends on external illumination for its visual effect (colour depending on the incident light). They have a life-time of about 50,000 hours.

LCDs have numerous uses of which a few are listed below;

- Field-effect LCDs are normally used in watches and portable instruments where source of energy is a prime consideration.
- Thousands of tiny LCDs are used to form the picture elements (pixels) of the screen in one type of B & W pocket TV receiver.
- Recent desktop LCD monitors and Notebook computer display

- Cellular phone display to display data on personal digital assistant (PDAs) such as Palm V_x etc

5.4 Military Uses of LCD Monitors

LCD monitors have been adopted by the United States of America military instead of CRT displays because they are smaller, lighter and more efficient, although monochrome plasma displays are also used, notably for their M1 Abrams tanks. For use with night vision imaging systems a US military LCD monitor must be compliant with MIL-L-3009 (formerly MIL-L-85762A). These LCD monitors go through extensive certification so that they pass the standards for the military. These include MIL-STD-901D - High Shock (Sea Vessels), MIL-STD-167B - Vibration (Sea Vessels), MIL-STD-810F – Field Environmental Conditions (Ground Vehicles and Systems), MIL-STD-461E/F – EMI/RFI (Electromagnetic Interference/Radio Frequency Interference), MIL-STD-740B – Airborne/Structure borne Noise, and TEMPEST - Telecommunications Electronics Material Protected from Emanating Spurious Transmissions.

6.0 Conclusion

LCDs have displaced cathode ray tube (CRT) displays in most applications. They are usually more compact, lightweight, portable, less expensive, more reliable, and easier on the eyes. They are available in a wider range of screen sizes than CRT and plasma displays, and since they do not use phosphors, they cannot suffer image burn-in.
LCDs are more energy efficient and offer safer disposal than CRTs. Its low electrical power consumption enables it to be used in battery-powered electronic equipment. It is an electronically-modulated optical device made up of any number of pixels filled with liquid crystals and arrayed in front of a light source (backlight) or reflector to produce images in colour or monochrome. The earliest discovery leading to the development of LCD technology, the discovery of liquid crystals, dates from 1888 (Steed and Atwood, 2009). By 2008, worldwide sales of televisions with LCD screens had surpassed the sale of CRT units.

7.0 References

7.1 Books and Article references

1. Alkeskjold T.T., *et al.* (2007). "Integrating liquid crystal based optical devices in photonic crystal". *Optical and Quantum Electronics* 39 (12–13): 1009. doi:10.1007/s11082-007-9139-8
2. Castellano A.J., (2005). *Liquid Gold: The Story of Liquid Crystal Displays and the Creation of an Industry*. World Scientific Publishing. ISBN 978-9812389565.
3. Castellano, A.J., (2006). "Modifying Light". *American Scientist* 94 (5): 438–445.
4. Chandrasekhar S., (1994), Liquid Crystals, Cambridge University Press
5. Collings, P.J. and Hird, M (1997). *Introduction to Liquid Crystals*. Bristol, PA: Taylor & Francis. ISBN 0-7484-0643-3.
6. GameBoy : User Manual, Page 12". Retrieved 2011-02-12.
7. Lee, K. H., Kim, H. Y., Park, K. H., Jang, S. J., Park, I. C. and Lee, J. Y., (June 2006). "A Novel Outdoor Readability of Portable TFT-LCD with AFFS Technology". *SID Symposium Digest of Technical Papers* (AIP) 37 (1): 1079–1082. doi:10.1889/1.2433159
8. Steed J.W. and Atwood J.L. (2009). *Supramolecular Chemistry* (2nd ed.). John Wiley and Sons. pp. 844. ISBN 9780470512340.
9. Uche C (PhD.) (2007): "Development of bistable displays". University of oxford. Retrieved 2007-07-18
10. Yang, D.K., Wu, S.T., (2006). Fundamentals of Liquid Crystal Devices, Wiley SID Series in Display Technology.

7.2 Web References

1. Anders Jacobsen's blog (January 4, 2006): "What is the ISO 13406-2 standard for LCD screen pixel faults?"
   http://www.jacobsen.no/anders/blog/archives/2006/01/04/what_is_the_iso_134062_standard for_lcd_screen_pixel_faults.html
2. Forbes.com (December 30, 2004): "Samsung to Offer 'Zero-PIXEL-DEFECT' Warranty for LCD Monitors":
   http://www.forbes.com/infoimaging/feeds/infoimaging/2004/12/30/infoimagingasiapulse_20 04_12_30_ix_9333-0197-.html
3. Lenovo (June 25, 2007): "Display (LCD) replacement for defective pixels - ThinkPad":
   http://www-307.ibm.com/pc/support/site.wss/document.do?lndocid=MIGR-4U9P53
4. Nemoptics.com (2009): http://www.nemoptic.com/content.php?section=technology
5. Samsung (February 5, 2005): "What is Samsung's Policy on dead pixels?"
   http://web.archive.org/web/20070304043758/http:/erms.samsungelectronics.com/customer/uk/jsp/faqs/faqs_view.jsp?SITE_ID=31&PG_ID=16&AT_ID=17628&PROD_SUB_ID=546
6. Wikipedia.org: "Liquid Crystal": http://en.wikipedia.org/wiki/Liquid_crystal
7. Wikipedia.org: "Liquid Crystal Display":
   http://en.wikipedia.org/wiki/Liquid_crystal_display